UNLOCK YOUR RESEARCH POTENTIAL THROUGH YOGA

RESEARCH SCHOLARS COMPANION

VIJAY RAJPUROHIT

Copyright © 2018

WWW.RESEARCHVOYAGE.COM

WHY I WROTE THIS BOOK

> The more fit you are
> The more you can
> pursue your vision
>
> ---- Robin Sharma

In more than 20 years of working with people in academics and research, I have come in contact with many individuals who have achieved an incredible degree of success in various fields, but have found themselves struggling in research. As a research scholar some of the problems they have shared with me may be familiar to you.

" I've set and met my career goals and I'm having tremendous research success. But it's cost me my eyes. I have severe eye stress hazy spots and blurry vision. I've had to ask myself -- is it worth it?"

"As I work continuously working on my desktop I am suffering from back pain and doctor has termed it as "Cervical Spondylosis", I am really worried."

"I am not getting new ideas for my research. This is really stopping my future work. I don't know how to get ideas?"

"I forget everything when I am presenting any concept or if someone asks me some questions I cant able to recall the concepts which actually I have studied."

"There's so much research do. And there's never enough time. I feel pressured and hassled all day, every day, seven days a week."

"With a positive mental attitude I tell myself I can do it. But I don't. After a few weeks, I fizzle. I just can't seem to keep a promise I make to myself."

This book will help in finding solutions to the above mentioned problems in a systematic way through the power of Yoga.

There are countless asanas and awesome meditation techniques that all have startling benefits. But the problem is the selection of specific asanas from which a researcher can gain great benefits within the limited time constraints. I have actually identified 08 crucial health factors that matter a lot for a researcher. By adopting the yogasanas explained in this book a researcher can demonstrate a speedy progress in his research career.

WHY YOU SHOULD READ THIS BOOK

This book will help you to unlock your research potential through the practice of YOGA. The yogasanas presented in this book will assist you

i. In coming out of the health problems such as sleeplessness, cervical spondylosis, stress, anxiety, depression etc. which are likely to occur during your journey as research scholar.

ii. In improving your memory power required to remember the crucial basic research concepts

iii. To improve right brain activity which will in turn develop your creativity that is required for research.

Once you start practicing the YOGA you will find a slow transition from your past to the future in an unimaginable way. You will start seeing the unfolding of hidden talent within you. You will think about your health and research problems in an entirely different way. You will develop new mental toughness that will take you through the intricacies of research which others find really difficult to overcome. You can start now to fulfill your hunger for research that is always fresh and meaningful and filled with extra ordinary contributions- in the field of research right to the end.

TABLE OF CONTENTS

Contents

Why I Wrote This Book..2

Why You Should Read This Book...................................5

CHAPTER 1. FROM A RESEARCHER'S DIARY................10

   ADVANTAGES OF TAKING UP Ph.D. 10

   THE OTHER SIDE OF THE STICK 11

   YOGA...14

CHAPTER 2. REDUCING EYE STRESS 20

   1.   PALMING...21

   2.   BLINKING 22

   3.   EYE ROTATION 23

   4.   TRATAKASANA 25

   5.   BHASTRIKA PRANAYAMA 26

CHAPTER 3. HOW TO HANDLE CERVICAL SPONDYLOSIS 28

   1.   DHANURASANA (BOW POSE) 29

   2.   BHUJANGASANA (COBRA POSE)..................30

   3.   MATSYASANA(FISH POSE) 31

   4.   MARJARIASANA(CAT POSE) 33

   5.   MAKARASANA(CROCODILE POSE)...............34

   6.   PRANAYAMA..36

CHAPTER 4. STIMULATING RIGHT BRAIN ACTIVITY THROUGH YOGA 38
   1. SIRSASANA 40
   2. MAYURASANA (PEACOCK POSE) 42
   3. PADMASANA (LOTUS POSE) 43
   4. VAJRASANA (DIAMOND POSE) 45
   5. KAPALBHATI 46

CHAPTER 5. YOGASANAS FOR BETTER SLEEP 48
   1. VAJRASANASTHA POORVA YOGA MUDRA 49
   2. BHRAMARI RECHAKA PRANAYAMA 51
   3. SAHAJA CHANDRABHEDAN PRANAYAMA 52
   4. HASTAPADASANA 54
   5. BALASANA (CHILD POSE) 55

CHAPTER 6. INCREASING MEMORY POWER 57
   1. KAKASANA – CROW POSE 57
   2. PADAHASTASANA-FORWARD BENDING POSE 59
   3. SARVANGASANA - SHOULDER STAND POSE 60
   4. PASCHIMOTTANASANA– SEATED FORWARD BEND POSE 61
   5. HALASANA -PLOW POSE 62

CHAPTER 7. REDUCING ANXIETY 64
   1. SUKHASANA (EASY POSE) 65
   2. PASCHIMOTTANASANA (SEATED FORWARD BEND POSE) 67
   3. ANANDA BALASANA (HAPPY BABY POSE) 68

    4. NADI SHODHANA IN PRANAYAMA (ALTERNATE NOSTRIL BREATHING POSE) 69

    5. VIPARITA KARANI (LEGS UP THE WALL POSE) 70

    6. MEDITATION TECHNIQUES 71

CHAPTER 8. REDUCING DEPRESSION 74

    1. UTTANASANA (STANDING FORWARD BEND POSE) 74

    2. BALASANA (CHILD POSE) 76

    3. SETU BANDHASANA (BRIDGE POSE) 77

    4. SUPTA BADDHA KONASANA (RECLINED BOUND ANGLE POSE) 78

    5. ANANDA BALASANA (HAPPY BABY POSE) 79

    6. MEDITATION TECHNIQUES 80

CHAPTER 9. REDUCING STRESS 83

    1. GARUDASANA (EAGLE POSE) 84

    2. UTTANASANA (STANDING FORWARD BEND POSE) 85

    3. BALASANA (CHILD POSE) 86

    4. VAJRASANA (THUNDERBOLT POSE) 87

    5. SUPTA BADDHA KONASANA (RECLINED BOUND ANGLE POSE) 88

    6. MEDITATION TECHNIQUES 89

ABOUT THE AUTHOR 92

CHAPTER 1. FROM A RESEARCHER'S DIARY

You might have chosen to take a look at this book because you are currently feeling besieged by your PhD, or perhaps may be planning to pursue Ph.D. Then you are at the right place.

ADVANTAGES OF TAKING UP Ph.D.

A PhD (Doctor of Philosophy) involves hard work, meticulous thinking skills and a lot of invested time. However, you receive many benefits and rewards for having a doctorate degree, whether it is for personal satisfaction, social status, better employment or for the sake of education and knowledge itself.

A Ph.D. also means that you have a lot of knowledge and information about your field. Your knowledge is not only theoretical, but also of practical use, and you can share it with others for problem-solving. You are an expert and specialist in your area of study, and your educated opinion will be highly valued among students, academicians, friends, colleagues and even the media.

THE OTHER SIDE OF THE STICK

The prevalence of mental health problems is higher in PhD students than in the highly educated general population, highly educated employees, and higher education students. Based on one study a PhD student is likely to develop mental health problems 2.4 times more than those in the general population with an undergraduate degree.

Results based on 12 mental health symptoms showed that 32% of PhD students are at risk of having or developing a common psychiatric disorder, especially depression. Why is it, you might be asking yourself, that being vivid, talented individuals and apparently facing a world of opportunities, are suffering constant self-doubt, depression, anxiety and burnout? From what I've seen, it doesn't matter much what discipline you belong to or which university you go to when it comes to developing never-ending despondency.

Doing PhD is similar to sitting in a driverless uncontrolled car without knowing where it will take you and how long it will take. The main lesson a PhD student must learn is how to find inner reserves of patience: this is a long haul and you will find that keeping up your spirits, momentum and health become as important as keeping your eye on the topic of study.

The potential for loosing your patience, losing your interest, losing your health, and even losing your mind is very high, evidence of which is the dropout rate among researchers and the considerable number who take longer than the formal completion period.

The specific field work requirements and psychological demands needed in Ph.D. are extremely higher as compared to getting basic degree. Researcher has to persistently keep working with either laptop/desktop constantly looking at the screen. This may lead to many physical and psychological issues at the earlier stage in their life. Above all as it is a secluded work where, a research scholar always feel beleaguered and frazzled. During the course of his/her research work, the researcher may come across few of the following warning signs which should not be overlooked.

1. Unable to concentrate and focus on work
2. Unable to get good sleep
3. Feeling of insecurity, failure and frustration
4. Mental and physical exertion
5. Feeling overwhelmed and confused

I don't want to paint a completely black picture or focus only on the difficulties here. There are students who manage to enjoy doing Ph.D. and come out with flying colours. So what and how they do it differently?

We need to identify those negative associations that have tainted the picture of Ph.D. over time, and then set them to one side. There are tools and techniques out there that can help us to restructure our research life and reshape our attitude. We simply need to identify what works best for us individually.

YOGA

I, asked around to see which tools and survival strategies had helped people to complete their research successfully. The answers were as wonderfully diverse as the individuals I talked to.

They varied from listening to music to meditation and yoga. Others found strength in praying God, or insights from TED Talks, while some stressed the role that family and sports played for their motivation and health.

During the course of interviews I met one of my professors who has guided a handful of research scholars.

He told he has discovered yoga as a powerful medicine to the fatigue he was experiencing while working toward his PhD. Yoga, which includes asanas or pranayam, is a vast discipline and represents an entire philosophy of existence. Interestingly, as time went on, he began to, look increasingly to the practice of yoga to achieve success in his research carrier. He also told me one of the most important tools to get through the issues during Ph.D. – FOCUS. Many of my interviewees later agreed to the fact that focusing one small problem or small objective at a time helped them to separate their feelings about the long term goal, Ph.D.

A wavering focus is one of the major causes for low productivity at the work place or low grades at school and college. It is this same attribute that students, professionals, entrepreneurs and home-makers alike have been trying to achieve to uplift the quality of their lives. While a lack of focus is the mother of all follies, achieving better concentration is not really that difficult when one knows where to look for help.

Of the various focus improvement options available today, I felt yoga and pranayama are probably some of the oldest and time-tested methods that can take you closer to your goal. Though it may seem that yoga is more of a physical workout technique, it works on the mental level too. It makes the body more flexible and improves the immunity level. It also alleviates stress, relieves the nerves and calms the mind.

Yoga is a great combination of body movements, breathing and meditation. It is a powerful therapy to deal with physical pains, stress, anxiety, fatigue and depression. It is a modern way to stay fit in a busy schedule and gain rejuvenation of body and mind. You can perform it anywhere according to the availability of the space and time. Yoga helps to gain mental peace, stress reduction and improve psychological performance of the person. Performing yoga will help a researcher to struggle with difficult situations.

Practicing yoga at least one hour daily will make a great impact on the mood of a researcher. One can notice the significant reduction in depression and anxiety as well. Stress is a hidden enemy which not only affects your mind but also affect functioning of the body. Peace of mind is a basic requirement of the researchers and that can be easily achieved with the help of yoga. Regular practice of yoga releases emotional burdens and improve the capacity to focus on the work. Through yoga you will learn the skill of balance between physical, mental and spiritual functioning.

According to the psychological research 'researchers having low stress can perform better than others'. It is already proved that stress and depression always make the negative impact on the academic performance of the students. While researcher goes through various challenging situations, yoga can help them to achieve improved confidence and peace of mind. Apart from mental stress researchers also face many physical problems including loss of sleep due to longer working hours, neck pain due to constant sitting for the long time headache due to continuous thinking and eye burns due to the long time computer work. Yoga can be a suitable cure for all these physical issues affecting fitness of the body.

To sum up top benefits of the yoga are

1. It improves concentration

2. It improves the psychological, mental and physical well being

3. Yoga reduces anxiety, stress and depression

4. It boosts self-confidence and improves memory power

5. It helps to calm your mind and alternately helps to gain peace of mind

6. Yoga provides the energy to struggle with various physical as well as mental traumas.

Process of Ph.D. is definitely time consuming and stressful but definitely worth of investing precious time and efforts. There are lots of hurdles you will face at every step going towards success. Handling the difficult situations wisely is an amazing skill. If simplest tool like yoga can help you to gain most demanding things in your life in current period like peace of mind, stress less life and fitness of the body, then it is always preferable to GO FOR IT.

CHAPTER 2. REDUCING EYE STRESS

Eye health is crucial for any researcher. Continuous usage of laptops sometimes leads to eye burns/eye stress which makes the researcher to halt his work. As a researcher, you spend most part of our day in front of the computer screen, staring at Word documents, excel sheets or YouTube videos. But in the process, you forget about one of the most important parts of your body, the eyes. The ones that itch from sheer exhaustion and burn tirelessly after the end of a really long day. Due to working on a computer throughout the day and studying for a long time, eyes do not just tire, but the effect of stress can be seen in your other works also.

You might not feel the need to exercise your eyes because there aren't any immediate symptoms or signs of weak or tired eyes but if you want to guard yourself from hazy spots and blurry vision in the future, then I suggest you exercise them as often as possible. One of the best ways to do so it to try yoga exercises for eyes. Yoga for eyes is a certain type of eye workout that can be done at any given time of the day, only for few minutes.

These yogasanas and meditation techniques will help you reduce eye burn and eye stress.

1. PALMING

Procedure: To do Palming, sit in a relaxed position. Rub your palms touching each other energetically until you sense the heat glowing from them. Put your palms on your closed eyes firmly and feel the heat spread.

Fig 2.1: palming

Benefits: Eye burn and eye stress can be healed with this exercise. It is a fast and easy way to relax, it improves blood circulation in eyes.

2. BLINKING

Procedure: Keep your eyes open and sit comfortably. Blink Your Eyes 10 times at a rapid pace. Close your eyes for 20 seconds. Put your focus on your breath. Repeat this exercise 5 times.

Fig 2.2 : Eye Blinking

Benefits: Reduces eye stress, refreshes your eyes.

3. EYE ROTATION

Procedure: To do this rotation, sit down in Padmasana with your head and spine erect. Lift the right fist with thumb facing upwards. Keep your elbow straight while doing. Focus your eyes on the thumb. Make a circle with the thumb, keeping the elbow straight. Repeat this exercise five times each in clockwise and anti-clockwise direction. Repeat the process with the left thumb. Close and rest the eyes and relax completely. Inhale while completing the upper arc of the circle. Exhale while completing the lower arc.

Fig 2.3: Eye Rotation

Benefits: It helps eyes to rest and protect them from disorders and diseases. This exercise is ideal for those who spend a lot of time in front of the Desktop/laptop.

Advantages: Continuously moving your eyes helps remain eye disorder at bay and improve vision.

4. TRATAKASANA

Method: Trataka means to continuously see a point for a predetermined time. Sit down at ease, either in Padmasana or Vajrasana. Position a candle at about two feet from where you are sitting. Light the candle and gaze at the flame without blinking. You can count numbers in your head to keep track of time and for your mind to not waver. Look as long as you can. The longer you do, the better.

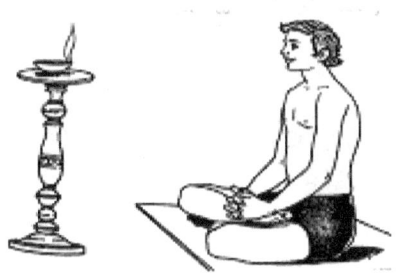

Fig 2.4: Tratakasana

Advantages: This improves concentration and vision and reduces stress on the eyes. Doing this improves your concentration and vision. This eye yogasana lowers high myopic eye power.

5. BHASTRIKA PRANAYAMA

Procedure: Sit in Padmasana with your spinal cord upright. Use your right thumb, close your right nostrils, inhale and exhale strongly and quickly through the left nostril. You can feel your abdominal walls roar while you do the exercise. Do it 25 times while exercising you can feel the walls of your stomach. Make your last breath long and deep. Now, repeat the same procedure on your right nostril with the left thumb closing the left nostril. Finishing the exercise in both nostrils makes a bhastrika. Relax for around 30 seconds and repeat the entire process again. Do it for about 10 minutes.

Fig 2.5 Bhastrika Pranayama

Benefits: Bhastrika pranayama increases blood circulation in the head and improves vision. It also refreshes your physical and mental health.

CHAPTER 3. HOW TO HANDLE CERVICAL SPONDYLOSIS

Today's modern world demands long hours of sitting in a particular posture at our work desk. These tiring long hours take a toll on both of our physical and mental conditions. The most common ailment that stems from these adversely affected physical and mental conditions is back pain caused due to abrasion and attrition of spine disks. This could occur even for the people of younger and middle ages. In middle parlance, such cervical and back problems are termed as "Cervical Spondylosis"

Regular practice of yoga is the most effective remedy for such physical problems as the "asanas" of yoga help to enhance the flexibility of body, calming of mind and growing perspective towards life.

These yogasanas and meditation techniques will help you to overcome Cervical Spondylosis.

1. DHANURASANA (BOW POSE)

Procedure: Lie down straight on the floor, exhale, bend your knees and hold the ankles with hands. While inhaling, raise thigh, chest, and head as per your comfort level. Try to maintain the body weight in the lower abdomen, join the ankles, look up and breathe normally. Now exhale slowly, bring down the head legs up to the knee joint. Maintain this pose as per comfort level and then slowly come back to original position.

Duration:20-30 seconds

Benefits: This particular bow poses to stretch and stimulate the neck. It is also good for maintaining weight, overcoming lethargy.

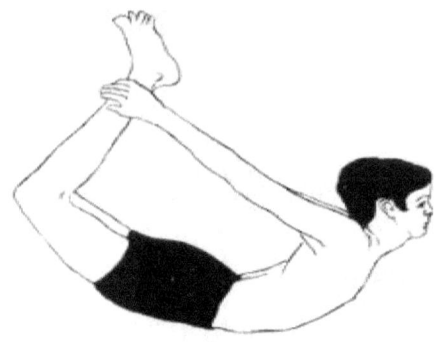

Fig 3.1 : Dhanurasana

Precaution: People who are suffering from a heart problem, high blood pressure should avoid this. Don't practice after having a meal. It is recommended not to do this asana before going to bed.

2. BHUJANGASANA (COBRA POSE)

Procedure: Bring your hands back underneath the shoulder, inhale and stretch up. You can curl your toes under. As you exhale, sit back, stretch your toes out. Sit back on your heels, stay there for a while and come to normal pose after few seconds.

Fig 3.2 : Bhujangasana

Duration: 30-40 seconds each

Benefits: This asana helps to remove stiffness of neck and shoulder and acts as spondylosis curer.

Precaution: Those who are suffering from stomach ulcers and pregnant woman are advised to avoid this particular asana. Those who have serious spinal injuries should also avoid this asana.

3. MATSYASANA(FISH POSE)

Procedure: This asana is going to relieve your neck muscle and going to relax as well. To do this, stretch both legs forward. Lie down on your back placing both the elbows right on the back. After that open your heart first as you inhale and slowly as you exhale drop the crown of the head all the way down to the ground.

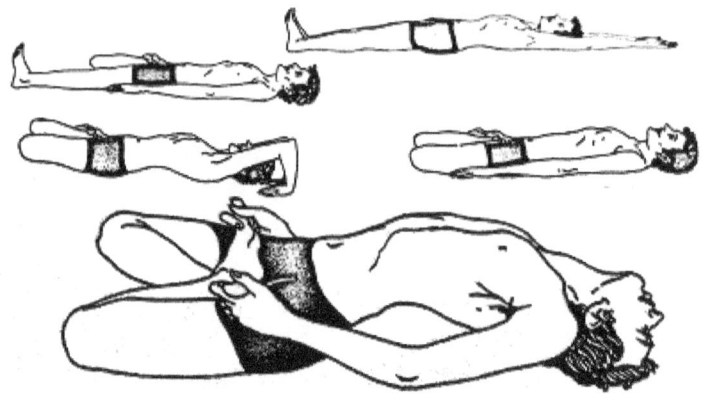

Fig 3.3 : Matsyasana

Duration:40-50 seconds each

Benefits: It provides strength and flexibility to the entire cervical region and helps in overcoming stress and strain caused due to long hour sitting.

Precaution: Patients with back injury, high blood pressure, insomnia and migraine should avoid this particular asana.

4. MARJARIASANA(CAT POSE)

Procedure: Kneel on the floor putting both palms on the floor pointing forward. Legs should be slightly apart and palm should be at shoulder length. Take a position of the cat. Relax your body. Now exhale completely and feel your belly go inwards. At the same time move head inward between shoulder. Now inhale and arch your back in opposite direction. The spine will bend downwards. The head, neck shoulder should be arched backward, as if looking up. Repeat this pose as many as you can.

Fig 3.4: Marjarasana

Duration: 60 seconds each

Benefits: This particular asana provides gentle massage to your spine and helps to loosen the vertebral column. Highly recommended for the people, who have a very rigid spine and chronic neck pain.

Precaution: People having neck injuries should avoid this pose. One should not do this pose beyond their comfort level. People who are suffering from back pain should avoid this asana.

5. MAKARASANA(CROCODILE POSE)

Procedure: Lie down on the floor on your stomach, fold the hands and keep the tip of elbows on the ground with your fingers facing upwards. Now raise your shoulder and neck. Keep your neck straight and look forward. Bend your head a little forward. Stretch your legs. Feel your body touching the ground. Breathe normally.

3.5 : Makarasana

Duration: 1-2 minute each

Benefits: This asana is good for the spine to resume its normal shape and very effective to release compression of the spinal nerves.

Precaution: Avoid moving your body or put stress on the body while doing this asana as it may disturb the practice. People suffering from exaggerated lumbar curve should avoid this asana.

6. PRANAYAMA

Procedure: Sit down cross-legged on the floor. Close your eyes and raise your right hand and place the elbow of your right hand on the palm of your left hand, bring the first finger of your right hand and the thumb together. Hold your breath and close the right nostril with your thumb. Exhale through your left nostril. Close your left nostril with the second and third fingers of your right hand and inhale deeply through the other. Repeat this pranayama for five minutes.

Fig 3.6: Pranayama

Duration:5 minutes

Benefits: This pranayama helps to cleanse your system and brings you health and vitality by regulating the vital life force in your body

CHAPTER 4. STIMULATING RIGHT BRAIN ACTIVITY THROUGH YOGA

A fully active right brain will help a researcher to take great steps towards fulfilling his goals. People with fully awakened righ brain have huge amount of energy which they can use for inventions, creation of complex products and solving challenging problems.

We can activate the right brain through a set of mental or physical activities. The brain needs proper nourishment, hydration and exercises as well just like any other body part. This will stimulate the right hemisphere of the brain to improve the thought process which is essential for a researcher.

Scientists have found out that our particular side body functions and movements are controlled by the opposite side of the brain. It is termed as 'alternative dominance of cerebral hemispheric activity'. As it turns out, the right hemisphere of the brain which impacts our creative nature controls the left side of our body. Conversely, the left hemisphere of the brain, the analytical part, controls the right side of our body. In simpler term, the right hemisphere of the brain gathers all the new experiences and sensations, and then the left hemisphere of the brain analyses and categories all those experiences into a cognitive whole.

Now, if any of these activities of the brain gets ineffective, then our body and mind loose balance and we fail to acquire accurate experience of the reality. With the help of yoga, we could restore this balance of logical and intuitive senses. The practices of yoga with its several "asana" and techniques of breathing help to maintain the rhythmical balance of the alternative cerebral activities in our body.

1. SIRSASANA

Procedure: This asana requires complete inversion of body and fine upper body strength. It is necessary that the stomach is empty and bowels clean. To do this asana first interlock the finger tightly and make a cup form with palm. Place the head on the formed cup so the crown of head touches the palm. Raise knees from the floor by pulling the toes towards the head. The spine and thighs should be in line and straight. Close the eyes and breathe deeply. Relax your body as much as you can.

Fig 4.1: Sirsasana

Duration: 1-5 minutes

Benefits: This asana instantly calms your mind, stimulates the pituitary gland, improves digestive power and increases the blood flow to the brain. It is extremely helpful to cure a migraine and headache.

Precaution: Try to put most of your weight of the body on the head not on arms. Persons suffering from high blood pressure, cerebral, slipped disc should avoid this asana.

2. MAYURASANA (PEACOCK POSE)

Procedure: This asana is best to do at morning when the stomach is empty. In this asana, you have to move your fingers towards your body, bend your elbows and press it towards the abdomen. Try to keep your belly firm and tight as you are pressing elbows on your abdomen. Now drop the head on the floor and work up straight on your stomach. Next stretch legs out so that knees are straight and upper part of the feet are facing the ground. Shift your body weight forward and lift your legs up.

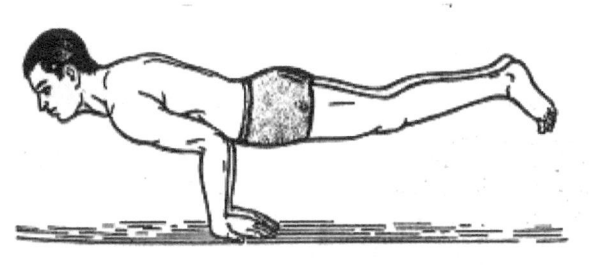

Fig 4.2: Mayurasana

Duration: 30-60 seconds

Benefits: This asana helps to improve concentration power and coordination between mind and body.

Precaution: Persons who are suffering from brain tumor, intestine problem and high blood pressure should avoid this asana. People who are having infection in the eyes, nose or ears should avoid this asana.

3. PADMASANA (LOTUS POSE)

Procedure: Sit with your legs stretched out, bend the right knee and place it on the left thigh. Repeat the same step with the other leg. Make sure that the soles of the feet point upwards and the heels are close to abdomen. Place your hands on your knees, close your eyes and relax.

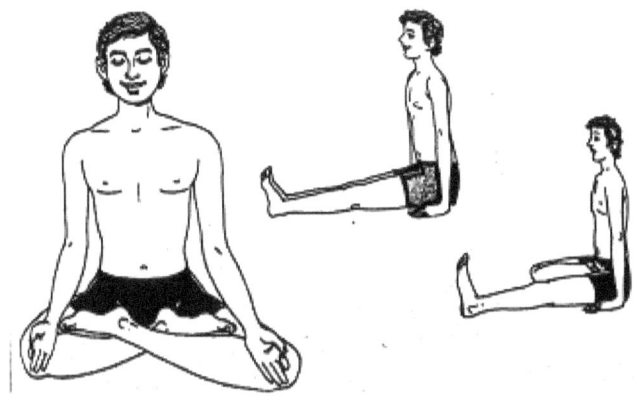

Fig 4.3: Padmasana

Duration: 1-5minutes

Benefit: Padmasana has a relaxing effect on nervous system. It helps to relax the mind and calms the brain. This particular asana awakens the chakras in your body and increases your awareness.

Precaution: If someone has injured or weak knees or sciatica should avoid this asana.

4. VAJRASANA (DIAMOND POSE)

Procedure: Kneel on the floor toes together and heels apart. Lower your buttocks to the inside surface of your feet. Place your hands on the knees, palms down. Close your eyes and relax.

Fig 4.4 : Vajrasama

Duration:5-10minutes

Benefit: This pose helps your body to relax and increase blood circulation in your brain.

Precaution: Person having joint pain should avoid this asana. People having spinal column ailments should not attempt this asana. Those who have a hernia, small or large intestine should avoid this.

5. KAPALBHATI

Fig 4.5: Kapalbhati

Procedure: Sit in padmasana. Push air forcefully out. The stomach will go in itself.

Duration: 5-10 minutes(30 time maximum in every 1 minute)

Benefit: Kapalbhati increases oxygen supply in your body, which stimulates and energizes the brain.

Precaution: As this pranayama is an advanced breathing technique, do not attempt it if you haven't practiced the basic pranayamas. Patients having high blood pressure, heart disease and pregnant woman should avoid this asana.

CHAPTER 5. YOGASANAS FOR BETTER SLEEP

In a life where you may always be on the move and not be able to find much time for yourself, things can get pretty stressful. This can lead to you not being able to get much sleep, or at least not a good night's sleep.

It may not fix your sleeping problems immediately, but it will be a slow progression that helps if you do it every single day before bed. You will get your body in a relaxed state and will have much less stress so your mind will be at ease.

One of the important lifestyle diseases of today's world is the sleep disorder. Lack of quality sleep or an improper sleeping pattern leads to several other ailments such as fatigue, anxiety, depression and others. Regular yoga practice acts as a non-invasive way to relieve the root cause of sleep disorder.

Sleep deprivation and stress can be a vicious cycle. We often have trouble falling sleep because we're worried and anxious, and in turn, the fact that we didn't get enough sleep makes us stressed the next day.

That's where yoga comes in. By lowering stress levels, calming the mind and relieving tension in the body, the soothing practice can be an effective natural sleep remedy. Certain resting and inversion poses can be particularly helpful for combatting restlessness and insomnia, especially when practiced in the evening or in bed before hitting the hay.

1. VAJRASANASTHA POORVA YOGA MUDRA

Procedure: First sit in vajrasana, then bring both the hands in front, place the palms on the ground, and then cross the hand, right palm on the left side and vice versa. Close them to each other. And then slowly rest the forehead on them. Adjust the length of the hand as per own convenience. Close your eyes. Allow your abdomen to relax. When you are coming out this state, first open your eyes, raise the head up and take the hand off the ground and again sit in vajrasana.

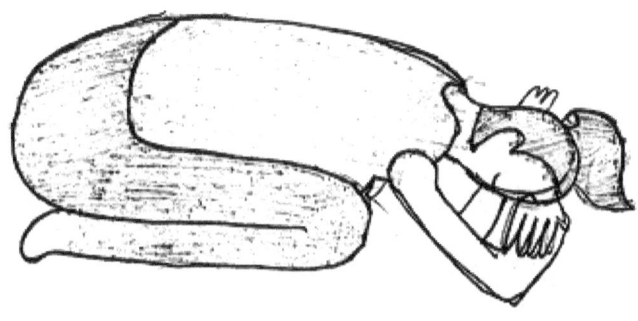

Fig 5.1 Vajrasanastha Poorva Yoga Mudra

Duration: 50-60 seconds

Benefits : This yoga will help you to relax the mind and relief the stress.

2. BHRAMARI RECHAKA PRANAYAMA

The *Bhramari pranayama* is a form of breathing which derives its name from the Indian bee Bhramari (Derived from Sanskrit Language). With this asana one can calm down his mind instantly. Common mental problems like agitation, frustration or anxiety can be handled effectively with this asana. This asana can is place and time independent, meaning it can be practiced anywhere - at work or home and is an instant option to de-stress yourself.

One of the step in this asana is to exhale with a **typical humming sound of a bee**, which explains why it is named so.

Procedure: Bhramari is a humming bee, and in this asana, one has to produce the sound of humming bee. To do this asana you have to sit in cross leg position. You have to put your right palm on the right eye and left palm on the left eye. . Then close the eyes and feel the body first. In this asana you have to inhale slowly to the full and while exhaling you chant the sound of "OM" without opening the mouth.

Fig 5.2 : Bhramari Rechaka Pranayama

Duration: 10-20 seconds each

Benefit: This pranayama immediately relax body and the mind

3. SAHAJA CHANDRABHEDAN PRANAYAMA

Procedure: Sit in cross leg position, left a hand on left knee with dhyana mudra and right hand on "ungli mudra", then put it in the right nostril. In this pranayama, the inhalation should always be with the left nose. After that close left nose and exhale through the right nose.

Duration: 1-2 seconds each

Fig 5.3 : Sahaja Chandrabhedana Pranayama

Benefit: This pranayama will help you to bring down your body temperature. To get the proper sleep it's important that your body temperature drops down. So this practice can be done just before going to sleep. You should do at least ten round before going to sleep.

4. HASTAPADASANA

Procedure: In this asana bring your hand to the feet. To do this asana you should have a comfortable distance between your feet. Pull your hand up, look at your fingers and then slowly bend your body and touch your feet

Duration:20-30 seconds

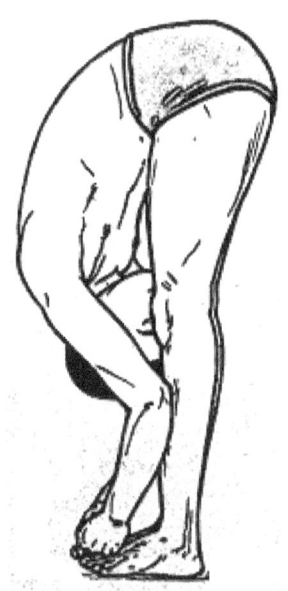

Fig 5.4 : Hastapadasana

Benefits: This asana helps to flow the blood towards the brain which helps to calm down the nerves system. You should do at least five times before sleep to get a better sleep.

Precaution: Those who are suffering from high blood pressure ulcer, cardiac problem and pregnant women should avoid this asana.

5. BALASANA (CHILD POSE)

Procedure: To do this you have to sit on your heels comfortably. Then roll your torso forward, bring your forehead to rest on the bed in front of you. After that lower your chest as close to your knees as comfortable as you can. Hold the pose and breath for one minute and then breathe.

Fig 5.5: Balasana

Duration: 1 minute each

Benefits: This particular pose is very helpful to release stress, relieve back and neck pain as well.

Precaution: it should not be performed if you are suffering from diarrhea, knee injury.

CHAPTER 6. INCREASING MEMORY POWER

Memory power is crucial for any researcher. A researcher with good memory power can link the already learnt basic concepts to new ideas and can achieve better progress in his research where as a researcher with poor memory may struggle to progress in his research.

However just like everything else, yoga and meditation techniques provide a concrete solution to transform your bad memory into good memory with these simple yogasanas and meditation techniques.

1. KAKASANA – CROW POSE

Procedure: Sit on the floor in a squatting position and keep your feet flat on the floor and an arm's length distance between the knees. Keep your palms in between the knees and place them firmly on the floor. Make sure to keep your elbows and knees at the same level. Next bend your torso forward and then lift legs up and try to balance your entire body on your palms. Keep your head in a straight position and look ahead.

Fig 6.1 : Kakasana

Duration: 30-60 seconds

Benefits: Kakasana helps improve concentration and coordination by maintaining a sense of balance.

Precautions: Avoid this yogasana if you are pregnant or are affected by chronic wrist pain or carpal tunnel syndrome.

2. PADAHASTASANA-FORWARD BENDING POSE

Procedure: Place your feet together and stand straight. Then start the pose by lifting your arms up over the head and touch the ears with it. Bend and reach for your feet. Your head and torso should be facing each other and hugging your thighs. Both your hands should be placed on either side of the feet.

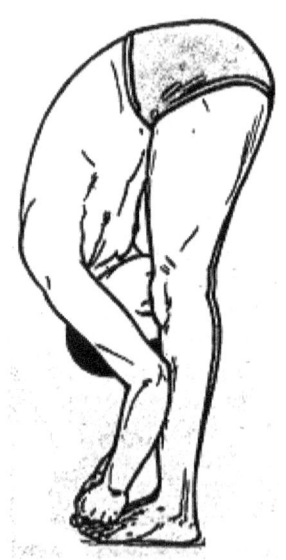

Fig 6.2: Padahastasana

Duration: 15-30 seconds

Benefits: Increases blood supply to the brain and invigorates the nervous system.

3. SARVANGASANA - SHOULDER STAND POSE

Procedure: Lie down on the back and keep your legs together. Keep your angles at a 90 degree angle by lifting them up. Place your arms against the ground, bend the elbows and hold your waist with hands. Then proceed to lift them up and take the legs higher to make a straight line. Make sure to keep your shoulder blades straight.

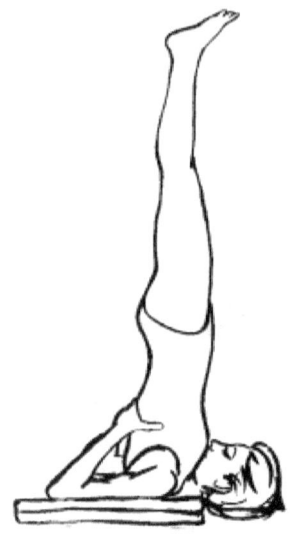

Fig 6.3: Sarvangasana

Duration: 30-60 seconds

Benefits: This pose relives hypertension, soothes headaches and cures insomnia.

4. PASCHIMOTTANASANA– SEATED FORWARD BEND POSE

Procedure: Sit down with your legs stretched out forward. Raise your hands straight up, with your arms touching the ears. Bend forward at the hips with your abdomen and chest hugging the thighs and your head on the knees. Your fingers should touch your toes, and you can keep your arms a little bent at the elbows.

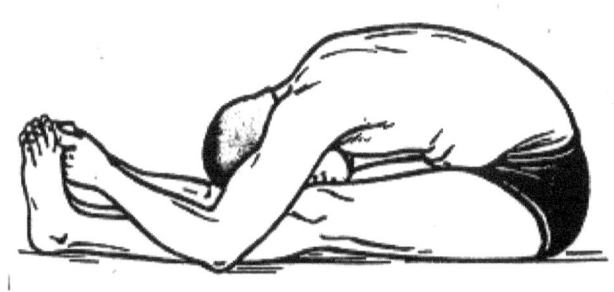

Fig 6.4: Paschimottasana

Duration: 30-60 seconds

Benefits: Increases concentration and cures headaches.

5. HALASANA - PLOW POSE

Duration: 30-60 seconds

Procedure: Lie on your back and keep arms on either side of the body with your palms facing downwards. Then life legs at a 90 degree angle. Next, support your hips with hands and lift them off the ground. Take the feet above your head at a 180 degree angle and touch your toes to the ground. Make sure to keep the back perpendicular to the ground. Bring back the hands to their initial position.

Fig 6.5 : Halasana

Benefits: This pose reduced fatigue and stress and calms your nervous system.

Precautions: Avoid doing this pose while pregnant, menstruating or while having neck pain or diarrhea.

CHAPTER 7. REDUCING ANXIETY

Anxiety is one's response to stress. These symptoms can be psychological physical or environmental challenges. There are various forms of anxiety include excessive worrying, a sense of fear, restlessness, overly emotional responses, and negative thinking. For some people, who are anxious may appear calm but the brain seems to be never quiet like cannot stop thinking. This situation can turn so bad and may interrupt the quality of life.

Regular yoga practice can help you stay calm and relaxed in daily life and can also give you the strength to face events as they come without getting restless. Yoga practice ideally includes the complete package of asanas (body postures), pranayamas (breathing techniques), meditation, helped several anxiety patients recover and face life with new positivity and the ancient yoga philosophy, all of which has and strength.

Every day we face several challenges and moments that worry us and stress our minds so much that we might end up having anxiety issues. Especially if you're a research scholar you need to take special care of your anxiety levels. Here are 5 Yogasanas and 5 meditation techniques which can help to reduce anxiety.

1. SUKHASANA (EASY POSE)

Procedure: Sit down on your spine and straighten your legs in front of you. Bend your knees and bring your feet together. Your palms should be on your knees while your back, neck and spine are aligned. Then gaze at whatever is in front of you and breathe deeply.

Fig 7.1 : Sukhasana

Duration: 50 – 60 seconds.
Benefits: It helps in calming you down to the level where anxiety is completely eliminated for the moment. It also relives you from mental worries.
Precautions: Avoid Sukhasana in case of recent hip injuries, spine disc problems and knee inflammations.

2. PASCHIMOTTANASANA (SEATED FORWARD BEND POSE)

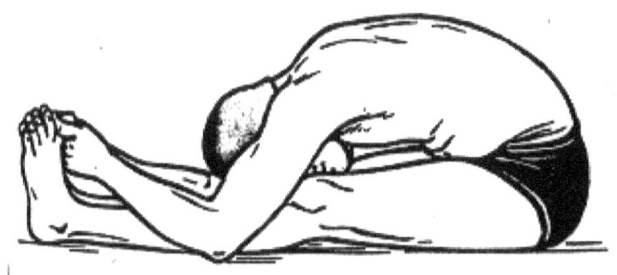

Fig 7.2 : Paschimottasana

Sit down on your mat and extend your feet forward. Slowly bend forward until your stomach touches the top of your thighs while you hold your feet with your hands.

Duration:20-30seconds.

Benefits: It helps you relieve stress, fatigue, and also fight with PMS, while also improving digestion. Precaution: Avoid Paschimottanasana in case of pregnancy, asthma, slipped disc problems and sciatica.

3. ANANDA BALASANA (HAPPY BABY POSE)

Lie down on your back and extend your legs and arms wide open. Bend your knees, bringing them in towards your belly. Stretch your hands and hold your feet.

Duration: 30 – 60 seconds.

Fig 7.3 : Anandabalasana

Benefits: It calms your mind and body. It relieves you from stress and fatigue every time you stretch your spine in this pose. Precautions: Avoid Ananda Balasana in case of neck or spine injuries, pregnancy and high blood pressure.

4. NADI SHODHANA IN PRANAYAMA (ALTERNATE NOSTRIL BREATHING POSE)

All you have to do is the swastikasana process in this yoga pose. Sit and relax. Focus on your breath then slowly take your right hand and place the forefinger and middle finger on your nose. First shut the right nostril with your right thumb and then do the same with your left. Take quick breaths. Duration:2-3 minutes.

Fig 7.4: Nadi Shodhana

Benefits: It balances the way the nervous system functions and improves concentration of the mind. Precautions: Avoid Nadi Shodhana in Pranayama if you're suffering from asthma and heart related problems.

5. VIPARITA KARANI (LEGS UP THE WALL POSE)

Procedure: Rest against a straight wall. Place your buttocks at the bottom of the wall and your legs on the wall. Relax your arms with the palms facing up. Close your eyes and focus on relaxing. Take deep breaths and hold the pose for as long as you can. Duration:5-10minutes

Fig 7.5: Viparita Karani

Benefits: It helps if you have headaches, and boosts energy in the body. It also helps relive back pain and cramps during menstruation. Precautions: Avoid Viparita Karani if you have hypertention and hernia.

6. MEDITATION TECHNIQUES

1. Pay Attention to Your Body: Your body needs your undivided attention especially when you're suffering from anxiety. When you focus on your body your mind tends to stop working. This relieves the pressure of the mind and at the same time relieves the body from stress and pain.

2. **Inhale and Exhale**: Take deep breaths. Deeply inhale and Exhale even more. When you exhale more the body tries to push away the stress. This also slows down your heart rate and you feel calm and relaxed.

3. **The Triangle Breathing**: After you inhale and exhale several times, and then start counting the number of breaths. If your long inhale takes 4 seconds, hold your breath for 4 seconds and exhale it until six seconds.

4. **Close Your Eyes and Exercise your Head and Neck**: Take a break from work and close your eyes. Breathe deeply, hold your breath and move your head from left to right and top to bottom. Do this slowly and breathe again. Straighten your back every 5 to 10 minutes.

5. **Do the Repetitive Prayer**: Repeat a short prayer in your mind while you close your eyes and focus

on the words you chant. Breathe in a uniform deep manner while you do this.

CHAPTER 8. REDUCING DEPRESSION

Depression is not a trivial matter to ignore. If you're a research scholar you mustn't be suffering from depression as it can hamper your health and career together. You can fight it by practicing simple yoga poses, and meditation at home. Here are few Yogasanas and meditation techniques which can help to reduce depression.

1. UTTANASANA (STANDING FORWARD BEND POSE)

Procedure: Stand straight, press your feet down to the ground and breathe powerfully. Gently bend down from the hips placing your stomach and chest on your thighs. Keep the weight on the heels of your feet.

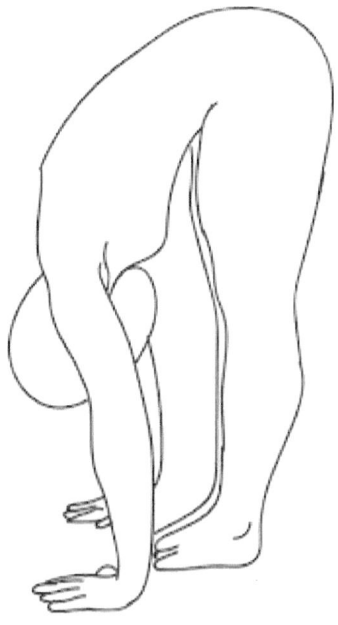

Fig 8.1: Uttanasana

Duration: 30 – 60 seconds
Benefits: Eases the tension caused due to stress in your shoulder, neck and back area.
Precautions: Avoid Uttanasana if you have had a back injury.

2. BALASANA (CHILD POSE)

Procedure: Sit down on your mat. Now bend forward to touch the ground and take your hands and lower back down to touch your toes. Fold your torso forward and let your eyebrows touch the mat.

Fig 8.2: Balasana

Duration: 2 – 3 minutes

Benefits: It sooths the mind and adrenals.
Precautions: Avoid Balasana in case of knee injury and pregnancy.

3. SETU BANDHASANA (BRIDGE POSE)

Procedure: Lie on your back. You back should be flat and your face upwards. Your hands should be straight on either side. Lift your hips upwards. Do this slowly while inhaling and bring your hip back down slowly while exhaling. **Duration**: 30 – 60 seconds.

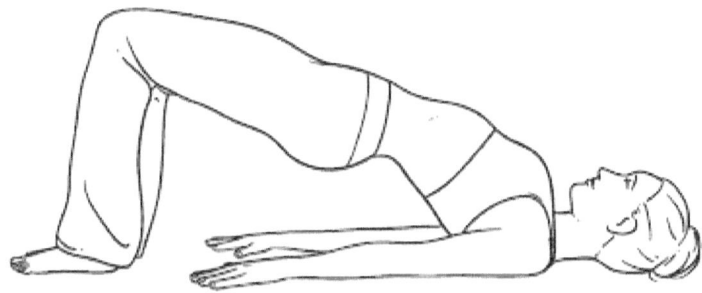

Fig 8.3: Setu Bandhasana

Benefits: It is great for reducing stress and fighting depression. It also helps in maintaining the blood pressure.

Precautions: Avoid Setu Bandhasana in case of neck pain, back injury, shoulder injury and knee pain.

4. SUPTA BADDHA KONASANA (RECLINED BOUND ANGLE POSE)

Procedure: Lie down on your back and bring your feet together. Keep your knees wide apart. Take your arms over your head and let them rest alongside your torso.

Fig 8.4 : Supta Baddha Konasana

Duration: 3 – 5 minutes

Benefits: It calms down and relaxes your mind and body

Precaution: Avoid Supta Baddha Konasana if you're suffering from shoulder or hip injury.

5. ANANDA BALASANA (HAPPY BABY POSE)

Procedure: Lie down on your back and extend your legs and arms wide open. Bend your knees, bringing them in towards your belly. Stretch your hands and hold your feet.

Duration: 30–60 seconds.

Benefits: It calms your mind and body. It relieves you from stress and fatigue every time you stretch your spine in this pose.

Fig 8.5 : Ananda Balasana

Precautions: Avoid Ananda Balasana in case of neck or spine injuries, pregnancy and high blood pressure.

6. MEDITATION TECHNIQUES

Boost Your Concentration: When you focus your attention on your subject or on any particular matter, your mind is not depressed and is functioning. You should keep your mind busy by trying not to think of sad or depressing things but instead try to resolve an issue or decipher a code of some kind. This is help you boost your concentration and focus on your research.

Diaphragmatic Breathing:
Take deep breaths. Breathe with from your belly, this will help you take longer and deeper breaths. Chest breathing is more like stressed breathing, which is not enough for our stressing minds and heart. Stress hormones are released when your heart rate is increased and blood pressure goes up. Try avoiding chest breathing to avoid depression.

Mindful Meditation:

In the process of mindful meditation, you do not stop thinking but you push aside the thoughts that come to your mind. In fact, you can listen to the thoughts that come to your mind and pay attention to them if they appear to be non-stressful and relaxing.

Guided Imagery:

This is one of the most popular techniques to help reduce anxiety and stress. It also helps in depression, high blood pressure, sleeping disorders and pain management. It uses the power of visualization to achieve a specific goal and improve performance. It also is good for research scholars as it enhances performance and develops skills.

Self Hypnosis:

This is a state in which your mind is extremely relaxed. You are forced to imagine something and focus on that one thing. It's good to put yourself in this state so that your mind is off the regular problems in life. This is very effective in cases of depression.

CHAPTER 9. REDUCING STRESS

Stress is an important issue and growing quickly in every facet of life. Stress is the process that occurs in response to events that dislocate, or threaten to disrupt, our physical or psychological performance.

Stress can induce feeling of frustration, fear, conflict, pressure, hurt, anger, sadness, inadequacy, guilt, loneliness or confusion. These problems can lead to academic/research failure, family conflicts, drug abuse etc.

Stress does appear to be a significant issue amongst PhD students, where a number of students appear to be suffering from serious physical symptoms of stress. The biggest causes of stress are Amount of work , deadlines for work, Uncertainty in the direction of your Lack of knowledge / skills to complete your work.

Here are five Yogasanas and meditation techniques which can help you to reduce stress.

1. GARUDASANA (EAGLE POSE)

Procedure: Stand straight and keep your arms wide apart. Now bring the right arm over the left arm and bend your elbows. Try to join your palms and bend your knees a little. Then hook your right feet behind your left calf. Engage your core and make sure you breathe effortlessly.

Fig 9.1 : Garudasana

Duration:10–20seconds

Benefits: It reduces the emotional tension and manages stress in an excellent manner.
Precautions: Avoid Garudasana if you had a recent knee, elbow or ankle injury.

2. UTTANASANA (STANDING FORWARD BEND POSE)

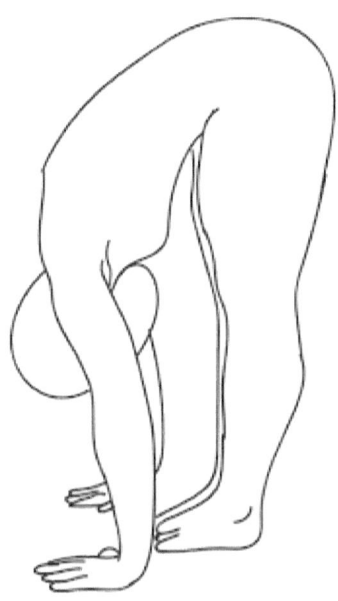

Fig 9.2: Uttansana

Procedure: Stand straight, press your feet down to the ground and breathe powerfully. Gently bend down from the hips placing your stomach and chest on your thighs. Keep the weight on the heels of your feet.

Duration: 30 – 60 seconds.

Benefits: Eases the tension caused due to stress in your shoulder, neck and back area. Precautions: Avoid Uttanasana if you have had a back injury.

3. BALASANA (CHILD POSE)

Procedure: Sit down on your mat. Now bend forward to touch the ground and take your hands and lower back down to touch your toes. Fold your torso forward and let your eyebrows touch the mat. Duration: 2 – 3 minutes

Fig 9.3: Balasana

Benefits: It sooths the mind and adrenals. Precautions: Avoid Balasana in case of knee injury and pregnancy.

4. VAJRASANA (THUNDERBOLT POSE)

Procedure: Sit back on your heels maintaining a length in the spine. Cross your hands in front of your chest while laying your palms within your underarms. You can sit cross legged or on your knees and feet.
Duration: 5- 10 minutes.

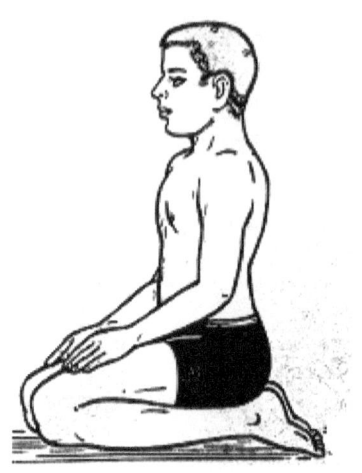

Fig 9.4: Vajrasana

Benefits: It gives you a restless sleep, so that your stress is reduced.

Precaution: Avoid Vajrasana if you're suffering from joint pain.

5. SUPTA BADDHA KONASANA (RECLINED BOUND ANGLE POSE)

Procedure: Lie down on your back and bring your feet together. Keep your knees wide apart. Take your arms over your head and let them rest alongside your torso.

Duration: 3 - 5 minutes

Benefits: It calms down and relaxes your mind and body

Fig 9.5 Supta Baddha Konasana

Precaution: Avoid Supta Baddha Konasana if you're suffering from shoulder or hip injury.

6. MEDITATION TECHNIQUES

Take Deep Breathe:

Stress forces you to take quick breaths. Your breaths mustn't be shallow, they should be deep and fulfilling. Your body requires more oxygen when it's stressed and much lesser when it is relaxed. So take few deep breaths until the body has relaxed. This will calm down your muscles and nerves eventually reducing stress and bringing a smile on your face.

Slow Down:

Slow down the pace of your work. No matter what you're doing, try to do it slowly and without pressurizing your mind too much. When your mind is under pressure, it works fasters and this increases you heart-rate while your emotions begin to outpour. This is you stressing yourself. Slowing down your work speed can help you a lot.

Relax Your Body:

It often happens that when you're deeply involved in your work, you tend to ignore your shoulder and back pain. Do not ignore these things, take a break to close your eyes and relax. Breathe in and out several times. Drink some cold water and relax for a minute. Your muscles and nerves will feel better.

Lift Your Head and Exercise It :

All the mental pressure is on your head, shoulders and spine. Lift your head and look upwards. Also try moving your head sideways, slowly from left to right, up to down and in circular motion. This will calm your mind and take it off the work eventually bringing down the stress level.

Straighten The Spine:

Straighten your spine whenever you need to. This instantly works to reduce your body stress and you feel better. Do this every 30 minutes. Close your eyes and breathe deeply while you do it. Pay attention to what your body tells you, especially when it's in pain. Try to relieve it off the pain by exercising it for 10 to 15 seconds.

ABOUT THE AUTHOR

Vijay Rajpurohit is an educator. Vijay loves educating and inspiring research scholars to succeed and live the life of their dreams. He has published a book on "Research Paper Writing: A Practical Approach" available at Amazon. He also writes blog posts for research scholars at www.researchvoyage.com

www.ingramcontent.com/pod-product-compliance
Lightning Source LLC
Chambersburg PA
CBHW030443220526
45464CB00006B/2402